~~ING~~ STEM
FROM
BASEBALL

How Does a Curveball Curve?
And Other Amazing Answers for Kids!

MARNE VENTURA

Table of Contents

Chapter Four: Math Makes Baseball More Fun

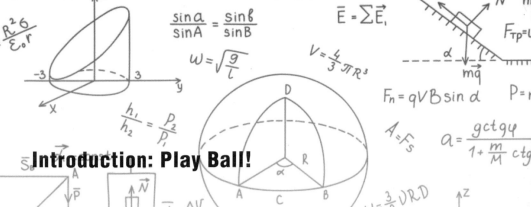

Introduction: Play Ball!

It's a sunny day at the ballpark. Fans sit side-by-side on the bleachers. On the field, the bases are loaded. The pitcher looks at each base from the mound. He nods to the catcher, winds up, and throws. The batter swings. WHACK! The ball soars across the field. The crowd jumps up and cheers as popcorn and sodas spill to the ground. It's a grand slam!

Americans have been playing organized baseball since 1846. Professionals, amateurs, and little leaguers play. Families play during picnics in the park. Neighborhood kids play in the schoolyard. Co-workers form teams and play on weekends. The first World Series was held in 1903. Every year, the American League and the National League compete to win the best of seven games. An average of nearly 14 million fans watch the games.[1]

Do you love baseball? Then you love science, technology, engineering, and math (STEM) too! Baseball is STEM in action. Science explains every pitch and swing. Athletes and fans use technology to improve their performance and enjoy the sport. Engineers find ways to keep players safe with new helmets and gloves. Major League Baseball (MLB) math experts analyze the statistics from every pitch, hit, and run.

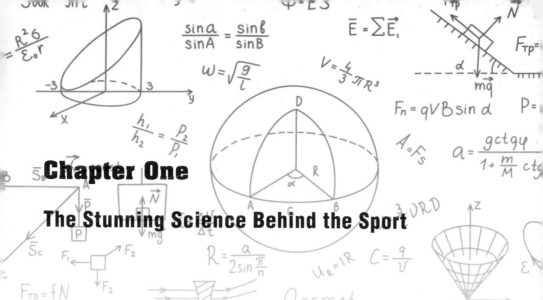

Chapter One

The Stunning Science Behind the Sport

Why do pitchers wind up?

What's that crazy dance before the pitch? Why not just throw the ball? Pitchers want the ball to be difficult to hit. A really fast ball achieves this goal. How do they do it? Physics!

Physics is the science of energy and objects in motion. Newton's Three Laws of Motion explain the way objects move. The first law states that every object will remain at rest or in motion in a straight line until acted upon by an outside force. The second law explains how the speed and direction of an object change when acted upon by an outside force. The third law states that for every action or force in nature, there is an equal and opposite reaction.[1]

According to the Second Law of Motion, the more force is put on an object, the faster it will move. The strength or force gained by motion is called **momentum**.[2] When a pitcher moves his body weight back behind the pitching mound, then thrusts it forward to throw the ball, he transfers momentum from his body to the ball. To get the fastest pitch, he starts by moving his largest muscles first. Then he continues through to his smallest muscles. In this way he drives his legs, then his hips, shoulders, arm, wrist, and fingers. Each movement of his body transfers momentum to the next movement.[3] As he takes a step forward and throws, he thrusts his whole body forward. An average MLB pitcher's ball will travel 95 miles per hour (153 km/h).[4]

Sir Isaac Newton

Sir Isaac Newton (1642–1727) was an English scientist and mathematician. His Three Laws of Motion became the basis for modern physics. Newton also explained gravity, the force that causes things like baseballs to fall toward the earth.[5]

How does a curveball curve?

Pitchers can make baseballs move in surprising ways. A curveball looks like it's speeding right over home plate. Then it suddenly drops and moves to one side. Pitchers throw curveballs to trick batters into swinging and missing.[6]

As soon as the ball leaves the pitcher's hand, the forces of nature affect its movement. Earth's **gravity** pulls it downward. Earth's air also has an effect on the ball. Air is made up of invisible gas **molecules**.[7] When these molecules hit the ball, they create **friction**, or drag. The bumps made by the stitches on the ball add to the drag and slow the ball down. Professional pitchers know how to throw pitches to take advantage of the forces of nature.

To throw a curveball, the pitcher snaps and turns his wrist. This gives the ball "topspin" as it flies through the air, which causes changes in the air molecules around it. The pull of gravity on air molecules gives them weight.[8] Fewer air molecules in an area have less weight, and thus lower air pressure. More air molecules in an area have more weight, and thus higher air pressure.[9] On the side of the ball that spins in the direction the ball is moving, the air pressure becomes higher. On the other side of the ball, the air pressure becomes lower. High-pressure air pushes toward low-pressure air. The difference in the air pressure on the sides of the ball causes the ball to curve.[10] This is called the Magnus effect.

Idioms

An idiom is a phrase used to mean something different than what the actual words express. The idiom "throw a curveball" comes from baseball. The phrase means to trick or surprise someone with something unexpected, usually causing trouble.[11-12]

Where is the sweet spot on a bat?

The batter steps up to plate. He touches the bat to the ground, then lifts it over his shoulder and bends his knees. The pitcher winds up and throws. The batter swings. WHACK! The ball flies out of the park. The crowd cheers as the batter runs around the bases and scores a home run.

One of the results of the collision between the bat and the ball is **vibration**. When the ball and the bat slam into each other, the bat shakes, or vibrates. Certain places on the bat shake more than others when they are hit by the ball. Over the years, batters have found that if they hit the ball at a certain place on the bat, it will cause the least amount of vibration. This spot is about six inches (15 cm) from the tip of the bat. When the bat shakes less, more energy is transferred from the bat to the ball. This makes the ball move faster and farther. Batters can identify this sweet spot because the collision makes a whack sound rather than a thud sound. And since the bat doesn't vibrate as much when the sweet spot hits the ball, it doesn't sting the batter's hands when he swings.[13]

Exit Velocity

The speed of a ball flying off the bat is called exit **velocity**. Yankees player Giancarlo Stanton broke the record for exit velocity when he hit a single off Red Sox pitcher Andrew Cashner in 2019. The ball was moving at 120.6 mph (194 km/h).[14–15]

Who is Tommy John?

Tommy John was a pitcher for the Los Angeles Dodgers. In 1974, his pitching arm gave out. His doctor, Frank Jobe, told him to rest and ice it for a few days. When it didn't get better, Dr. Jobe told John that he would need surgery to repair the band that connected his upper arm to his forearm, called a Ulnar Collateral Ligament (UCL).

When Jobe operated on John, he discovered that there was no band to repair. John's UCL was completely worn away from pitching so much. Dr. Jobe tried a new approach. He removed some healthy band from John's right wrist and attached it to John's elbow.[16] John started to use his elbow in hopes of being able to pitch again. In 1976, John returned to Major League Baseball. He went on to throw 207 innings. John won 164 games after his surgery.

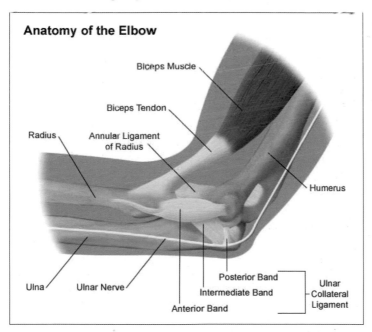

Anatomy of the Elbow

Biceps Muscle

Biceps Tendon

Radius

Annular Ligament of Radius

Humerus

Ulna

Ulnar Nerve

Posterior Band

Intermediate Band

Anterior Band

Ulnar Collateral Ligament

Now called Tommy John (TJ) surgery, Dr. Jobe's new procedure had a huge impact on MLB pitchers. Before TJ surgery, a torn UCL meant a pitcher's career was over. Since Tommy John's return to MLB in 1976, dozens of pitchers have had the surgery and continued their careers as pitchers.

Frank Wilson Jobe

Dr. Jobe (1925-2014) has been called "the godfather of sports medicine" for finding new ways to repair sports injuries. He was honored at a ceremony at the Baseball Hall of Fame in Cooperstown, NY, in 2013.[17]

Tommy John (left); Dr. Frank Jobe (right)

Do athletes eat ballpark food?

Ballpark snacks are a big part of the MLB experience. Have you ever wondered why ballparks sell hot dogs, nachos, soda, caramel-covered popcorn, and cotton candy to fans? Is that what ballplayers eat too?

Most modern MLB players use new information about nutrition in order to eat a healthy diet. Some teams hire a full-time registered dietitian to help players design their meals. Their goal is to use the science of nutrition to make sure players perform their best. Some players monitor their levels of vitamins to make sure that they're eating enough of the right foods. In addition to improving performance, they hope to prevent injury.

So, what do MLB players eat? Most follow the basic rules of healthy eating. They include plenty of fresh vegetables and fruits. They choose whole grains more often than refined grains. They get protein from sources such as fish, lean meats,

poultry, eggs, nuts, and beans. They drink plenty of water. They have treats on special occasions, but not every day.[18]

Most ballpark concession stand food is not what MLB players eat for top performance. So why do they sell it? Like so many things in baseball, it's a tradition that spectators enjoy.

Outrageous Snacks

Ballpark concession stands are known for serving crazy treats. Chase Field in Arizona serves a Churro Dog. It's a churro, wrapped in a chocolate-glazed donut, topped with frozen yogurt and caramel sauce.[19]

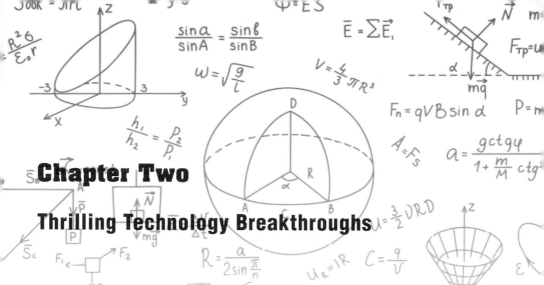

Chapter Two

Thrilling Technology Breakthroughs

How do players use radar?

Radar stands for **ra**dio **d**etection **a**nd **r**anging. A radar device bounces radio waves off of an object in order to tell where and how far away the object is, based on how long it takes for the waves to bounce back.[1]

The handheld radar guns used in baseball measure more than distance or location. They use a phenomenon called the Doppler effect to measure speed as well. Have you ever

noticed that the tone of an ambulance siren sounds higher as it comes close to your location, and lower as it drives away? The sound waves from the approaching siren are squeezed into a shorter distance. Because they're closer together, they have a higher **frequency** and a higher tone. The opposite thing happens as the siren moves away.[2] In the same way, when a pitch is moving toward a radar gun, the frequency of the bounced waves increases. Sensors inside the gun detect the frequency. A computer in the gun uses this **data** to compute the speed of the pitch.[3]

Some experts believe that the feedback that players get by measuring the speed of their pitch helps them improve. Today's pitchers are throwing harder and faster than ever before.

Feedback for Fans

Radar measures the speed of every pitch during an MLB game. The number is posted on the ballpark scoreboard as soon as it's thrown. For people watching at home, the speed of the pitch is shown on the television screen.[4]

When was the pitching machine invented?

Charles Hinton was a mathematician at Princeton University in the 1890s. While watching the Princeton baseball team practice, he saw a problem. Pitchers were wearing out their arms so that batters could practice hitting the balls. The pitchers risked fatigue and injury. Hinton decided to look for a solution.

First, he built a catapult. The aim was way off. Next he tried a cannon. He attached a rifle loaded with blanks to the cannon. The user shot the blanks into the back end of the cannon which heated up the gases inside the cannon. The hot gases, in turn, forced the baseball out of the cannon. Hinton's machine was not a big success. Players were afraid of the gunpowder explosions. [5-6]

In 1952, mechanical engineer Paul Giovagnoli came up with a better idea. He built "Iron Mike" using junkyard scraps.[7] Iron Mike has a shoulder, arm, hand, and finger. It is powered by a motor and a spring. The motor is joined to the shoulder with belts, pulleys, gears, and a drive chain. When a gear turns, it pushes on the shoulder. This moves the arm and pulls a cable that joins the shoulder to the spring. The spring tightens until the arm snaps forward to throw the pitch.[8]

Modern Machines

Today, pitching machines use wheels, mechanical arms, or pressurized air to throw balls to batters. Some can throw different kinds of pitches such as knuckleballs. The user controls the speed. Some machines project a video of a pitcher winding up and throwing.[9]

Why do players watch video recordings of themselves in action?

Tony Gwynn played for the San Diego Padres for 20 seasons, starting in 1982. He was one of the greatest hitters in baseball history. He was known for using video recordings to study and improve his swing.[10] He would record the TV broadcast of his games and then at home later, he would watch himself batting to try to learn from his hits as well as from his misses.[11]

The use of video to help baseball players become better hitters has come a long way since the 1980s. Today, many MLB players use high speed cameras that can shoot 22,000 frames per second while they practice. By watching a slow-motion recording of a pitch, players and coaches can see beyond what they are able to see with their eyes alone.[12] They might see how a pitcher's grip changes as he lets go of the ball. They might see how tiny changes in finger position affect the way the ball rotates.[13] They use the information to instruct players to do more of what works, and to stop doing what doesn't work. Video technology has helped modern pitchers throw harder, faster, and better.[14]

Tony Gwynn (1960-2014)

Right fielder for the San Diego Padres, Tony Gwynn, set the National League (NL) record for most consecutive seasons hitting .300 or better. He tied the NL record for most batting titles. He was the 22nd player to reach 3,000 hits.[16]

What's virtual reality?

In computer technology, virtual means not existing in the real, physical world, but instead being made by software to simulate reality.[16] Virtual reality (VR) is a computer-generated space that seems real. Baseball players use VR technology to practice their batting.[17]

A player puts on a headset with goggles. Looking through the goggles, he feels as if he is standing at home plate. In the distance he can see the stadium and scoreboard. It's like being inside of a video game that takes place at a ballpark.[18]

The player is the batter, and the computer system is the pitcher. To practice, the player holds a small bat that's also connected to the system. The computer is programmed to copy the pitches of top players. It throws pitches in every part of the strike zone. It randomly throws fastballs, curves, sliders, and change-ups.[19] The virtual pitcher winds up and throws to the player. The player presses a toggle when he wants to swing.[20]

The VR tool gives the batter feedback on the timing of his swing. Using virtual reality helps batters get better at knowing where the strike zone is located. They also improve at recognizing different pitches.[21]

Augmented Reality

When a computer system adds virtual reality to a real place, it's called "augmented reality."[22] One such app lets fans at a ballpark point their smartphone camera toward a player. The athlete's picture and data pop up on the fans' phones.[23]

What's wearable technology?

Just before the 2016 season, MLB approved the use of wearable technology during games. These devices are worn on different parts of the players' body.[24]

One device is a tightly fitting sleeve that the pitcher wears under his uniform. It covers his throwing arm from above the wrist to below the shoulder. It has a **sensor** inside of a slot, about the size of a quarter. The sensor sits on the inner forearm. When the player pitches, the sensor measures how fast the forearm rotates. The measurements are sent to a computer. Players use this feedback to improve their pitching.[25] The system also measures the fatigue or twisting force on a pitcher's arm. This can prevent injury by alerting coaches and players when the pitcher needs to rest.[26]

Wearable sensors are also used to help players master batting skills. A vest with sensors can track how a hitter's hands, batting arm, torso, and pelvis move as he hits. By keeping track of the data, trainers can see which types of movement work best.[27]

MLB rules don't allow players to get the data from these devices during the game, but they can study the data afterwards to improve for the next game.

Concussion Monitors

A hard blow to the head can be a dangerous injury. Therefore, researchers are working on a special mouth guard to monitor players' safety. It collects and analyzes saliva to tell if a player has had a concussion. This would enable players to get medical attention quickly.[28]

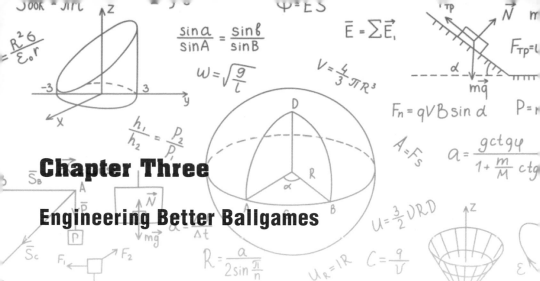

Chapter Three

Engineering Better Ballgames

When did ballplayers start wearing helmets?

Major League Baseball began in 1903 when the National League (NL) and the American League (AL) merged.[1] In the early days of baseball, most players did not wear anything to protect their heads during a game. In 1920, Ray Chapman, a batter for the Cleveland Indians, was struck in the head and

killed by a pitch from New York Yankee Carl Mays.[2] Despite this tragedy, protective helmets were still not commonly used for many more years.

In 1941, two Brooklyn Dodgers—Joe Medwick and Pee Wee Reese—were badly hit, or beaned, by pitches. Manager Larry MacPhail made the whole team wear protective helmets when batting. They were similar to baseball caps, but with a hard liner. It wasn't until 1971, more than fifty years after

Ray Chapman

Chapman's death, that MLB required all players to wear helmets. Twelve years later, in 1983, MLB made it mandatory for batters to wear helmets with flaps that cover the ear facing the pitcher.

In 2013, a new model of helmet was introduced. It's made of carbon fiber, the same material used to build spaceships. The new helmets are light, but strong enough to hold up against a 100 mph fastball.[3]

C-Flap Helmets

Some MLB players wear C-Flap helmets to protect their faces as well as their heads. A piece of protective plastic attaches to the earflap of the batting helmet. It covers the hitter's cheek and jaw.[4]

Why are baseball gloves webbed?

In the early days of baseball, players didn't wear gloves in the outfield. As more players were injured from playing bare-handed, gloves became more common. By the 1890s, most players wore padded leather gloves.[5]

In 1920, pitcher Bill Doak came up with a new glove design. It had straps between the thumb and index finger.[6] The idea of straps, or webbing, was a breakthrough. Players started catching balls in the web instead of the palm of their hand. It made the area where a ball could be caught much bigger. This increased the chances of catching the ball.[5] Today, baseball gloves all have webbing.

Glove designers vary the type of webbing based on the player's position on the field. For outfielders, big spaces between the straps let players see the ball as they catch it. Infield gloves have closely woven webbing. This protects their hands when they catch fast hard balls. First base gloves have no finger divisions, with open webbing that helps them grab and throw the ball quickly. Pitcher gloves have a web with no spaces. This lets them hide the ball, as well as their throwing hand, while they pitch. Catcher's mitts have heavy padding and closed webbing for catching fastballs.[8]

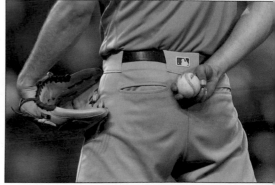

Not Tough Enough?

Why didn't the first ballplayers wear gloves? Much of the throwing was underhand, so balls were not thrown as hard and fast as they are today. Also, most players thought wearing a glove would make them look unmanly.[9]

What's a smart bat?

Maybe you've used a smartphone, or ridden in a smart car. Smart technology can do things that a computer does, often automatically. Smart devices are **electronic** tools that can connect, share, and interact with people and other devices.[10]

In recent years, engineers have come up with smart bats to help baseball players train. Designers start with a standard wood baseball bat with a wide barrel, narrow handle, and circular knob at the base. To make the bat "smart," they add a sensor that screws into the handle. The sensor connects wirelessly to an app on a smartphone. As the batter swings, the sensor measures the bat speed. It also measures the time when the bat hits the ball and tells the angle of the swing. The app has been programmed with information from batting experts. After the player swings, the app makes suggestions to the player on how to improve.[11–13] For example, if the batter is swinging down on the ball, it

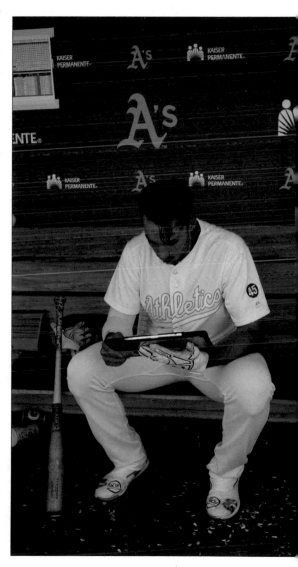

might be hitting the ground too soon. An uppercut angle without enough speed might mean easy-out pop ups. MLB allows the use of smart bats for training, but not during a game.[14]

Smart Coach

Data from a smart bat are transferred to a virtual coach in the app. If a batter has weak power, the app shows a video to teach the batter how to swing with more power.[15]

How are baseballs made?

The Rawlings Factory in Costa Rica makes all of the baseballs used in MLB games. Each ball has three layers. The center core is a sphere of cork coated with two layers of rubber. This is called the pill.[16] Next, four layers of gray sheep's wool yarn from New Zealand is wrapped around the pill. Then a layer of white wool yarn is wrapped around the gray yarn. Next is a layer of thinner gray wool yarn, and a layer of white cotton and polyester yarn. Finally, two pieces of cowhide from Tennessee are glued onto the ball, and sewn in place with red waxy thread.

MLB has strict rules for the size, shape, and weight of official baseballs. They must weigh between 5.0 and 5.25 ounces (.14–.15 kg). They must measure 2.86 inches to 2.94 inches (7.25–7.5 cm) in **diameter**. Their **circumference** must be between 9 inches and 9.25 inches (22.9–23.5 cm). The red

thread used to sew on the cowhide cover must be 88 inches (224 cm) long. There are always 108 stitches.[17]

For years, engineers worked to come up with a machine that would stitch the cover onto a baseball. They never succeeded, so today every official MLB baseball cover is still stitched by hand.[18]

Old Time Baseballs

The first players made their balls by winding yarn around a rubber core, and stitching on four sections of leather. In the 1870s, the Spalding Company came up with the idea of a cork center and a two-part cowhide cover.[19]

Can batters in MLB use any bat they want?

In the early days of baseball, players made their own bats. They were flat, or round, or even made from an axe handle.[20] Most were longer and heavier than the bats used today. Players used whatever kind of wood was available. In the 1870s, ash and maple wood became the most popular wood for bats.[21]

In 1884, a young fan named Bud Hillerich saw Pete Browning, a local baseball star, break his bat during a game in Louisville, Kentucky. After the game, Hillerich offered to make Browning a new bat in his father's woodworking shop. Together they made a bat out of ash wood. The bat worked so well that Hillerich and his father made more. They called the bat the "Louisville Slugger." This started the trend of players buying bats made by professionals.[22]

Around 1900, professional baseball rulemakers said bats had to be round, no more than 2.75 inches (7 cm) in diameter, no more than 42 inches (107 cm) long, and made only of hardwood. An 18 inch (46 cm) long section of the handle could be wrapped with twine.[23] Today MLB rules are the same as those set over a century ago.[24]

Aluminum Bats

In the 1970s, bat makers tried using aluminum instead of wood. When batters hit with aluminum bats, a phenomenon called the "trampoline effect" occurs. Because the ball bounces off with an unsafe amount of exit velocity, MLB chose not to allow aluminum bats.[25]

Chapter Four

Math Makes Baseball More Fun

Who was Henry Chadwick?

Henry Chadwick is often called the father of baseball. Chadwick was a sports reporter for a newspaper in New York. He loved the game of baseball, and helped to make it the popular game it is today. How did he do it? He used numbers.

Henry Chadwick was born in England in 1824. As a child he played rounders and cricket, two English games similar to baseball. In 1837 his family moved to New York. He became a baseball fan as a young sports reporter for *The New York Times*. He started ed writing about the games for newspapers.[1-2]

In the 1850s, people didn't have televisions or radios. Even photography was in its early stages. The only way to enjoy a game was to go to the ballpark, or read about it later in the newspaper. Chadwick came up with the box score. This was a grid where he

recorded the number of runs, hits, outs, assists, and errors for each player and team. Chadwick introduced batting average (AVG) as a tool to measure a hitter's success at the plate.[3] He also came up with the idea of earned and unearned runs. Box scores let fans enjoy the games from home. It was a big reason baseball became so popular in America.[4]

Why "K" for Strikeout?

Because Henry Chadwick's box score grid had to fit in a small space in a newspaper, he came up with letters to stand for baseball terms. "E" stood for error, "HR" was for home run, and "DP" was for double play. He couldn't use "S" for strikeout because he had already used it for sacrifice. At the time, *struck* was the popular term for strikeout. So, he used the last letter, "K."[5]

What is sabermetrics?

Sabermetrics is the way baseball teams and their fans use math to study baseball. SABR is short for the **S**ociety for **A**merican **B**aseball **R**esearch. It's a group of over 6,000 baseball fans around the world. Metrics is a method of measuring performance by using numbers.

American baseball writer Bill James gave sabermetrics its name.[6–7] In 1977, James began publishing a yearly booklet called *Baseball Abstract.* It was filled with math studies he did using numbers from box scores and other baseball records. James came up with new ways of interpreting a player's strength.[8] In the past, scouts ranked players based on their batting average, home runs, and runs batted in. James thought these numbers were not enough. They didn't count how many times the players walked, got on base, and avoided

outs.[9] James looked for ways to measure the player's worth to their team. For example, he created "Win Shares." This statistic helps teams rank a player's ability to help his team win games. James also came up with "Runs Created." This formula measures how much a batter helps the team score runs.[10]

Today, James is the world's most famous baseball analyst. From 2002 to 2019, he was an analyst for the Boston Red Sox.[11]

Sabermetrics at the Movies

Michael Lewis's book *Moneyball* came out in 2003. It told the story of how Oakland A's general manager Billy Beane used sabermetrics to hire players on a small budget. Brad Pitt starred as Billy Beane in the movie in 2011, while Jonah Hill played his assistant general manager.[13]

What is The Triple Crown award?

Baseball is a game of mathematics. Baseball teams and fans collect numbers for almost everything that happens during a game. These are **statistics**, or stats for short. Experts do math with the numbers to decide which players and teams did the best. They predict which team will win or lose the next game.[13]

Three important stats for batters are Batting Average (AVG), Home Runs (HRs), and Runs Batted In (RBIs). To find a player's batting average, divide the player's hits by his total at-bats. The result is a number between zero (shown as .000) and one (1.000).[14] For example, if a player gets 150 hits in a season and has 550 at bats, his batting average would be 150/550, or .273. Home runs is simply the number of home runs hit during a season. Runs batted in are earned when a batter hits the ball which allows another player to get back to home plate and score during the same play.[15]

There are two leagues that make up MLB: the American League (AL) and the National League (NL). If a batter has the top AVG, HRs, and RBIs for his league at the end of a season, he wins the batting Triple Crown.[16] Since 1920, only 10 players have won the Triple Crown award.[17]

Best of the Best

What kind of numbers does it take to win a Triple Crown? Mickey Mantle had 52 home runs in 1956. Lou Gehrig had 165 runs batted in during 1934. Roger Hornsby had the best batting average—.403—in 1925.[18]

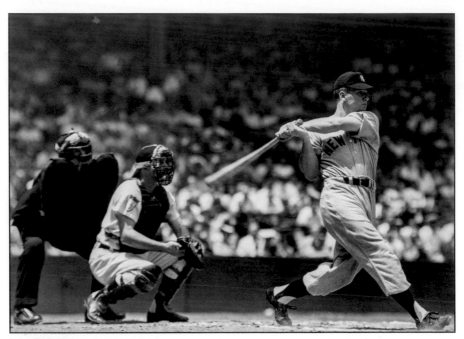

Lou Gehrig

Mickey Mantle

Roger Hornsby

Why do baseball fans love stats?

Numbers give fans a way to enjoy the game. Fans use stats to compare players and track their favorites.

For hitters, the most popular stats are batting average, home runs, runs batted in, runs scored, and stolen bases.[19] Today, some fans use on-base percentage (OBP) instead of batting average. OBP is how often a batter reaches base for each at-bat.[20]

Babe Ruth

For pitchers, the most popular stats are wins, earned run average (ERA), walks plus hits per inning pitched (WHIP), strikeouts, and saves.[21] In recent times, fans have used quality starts (QS) instead of wins to rank pitchers. A starting pitch-

Christy Mathewson

er gets a QS when he pitches at least six innings, and allows three or less earned runs.[22]

Some fans use stats to create their own fantasy baseball team. To play, fans join a league either online or in person. Each member of the league picks actual MLB players from multiple different professional teams for their team. During the MLB games, they keep track of their players' stats. The fantasy team with the most points at end of the season is the winner.[23]

Baseball Cards

Fans have been collecting and trading baseball cards since the 1860s.[24] Cards are printed with a picture of a player or a team on the front. The player's or team's statistics are printed on the back.[25]

Who does the math in baseball?

Mathematicians are some of the most important people in MLB. Radar and cameras in every MLB park record data for every game. To use these statistics, teams hire baseball analysts.[26]

Baseball analysts usually have college degrees in math, statistics, or computer science.[27] They study results from games and make predictions about future games. They study team and player performance. Additionally, they use computer software to find ways for the team to win more games. Analysts prepare reports for newspapers and television. They also help owners and managers decide which position players should take, who to hire, and how to rank players.[28-30]

Colleges now offer courses in baseball analytics. At Pomona College in California, students learn to collect data from special equipment. For example, one device measures velocity and spin rate from a pitching machine and sends it to an iPad. Students learn to study the numbers to make suggestions for better pitching. Another device is a bat sensor that records a hitter's swing path. By studying this data, students learn to identify the best angle for the swing.[31]

America's Pastime

What are the statistics for how much Americans love baseball? Nearly 170 million people said they were MLB fans in 2019. More than 68 million people went to an MLB game.[32] 23 million viewers watched the final 2019 World Series game on TV.[33]

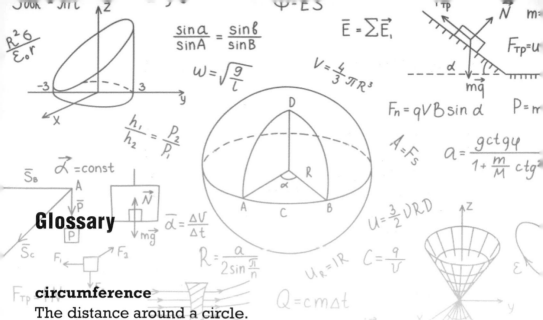

Glossary

circumference
The distance around a circle.

data
Numbers used to predict, understand, or plan.

diameter
A straight line passing from side to side through the center of a circle.

electronic
Using a computer to work.

frequency
The number of waves that pass a certain point in a certain time.

friction
The force that slows motion when two things come into contact.

gravity
The force that pulls an object toward the earth because of its mass.

molecule
The smallest particle of a substance.

momentum
The movement of an object, caused by its mass and motion.

radar
A device that sends out radio waves in order to locate an object.

sensor
A device that senses something and sends a signal to report it.

statistics
The collection of data in the form of numbers.

velocity
The speed of a moving object.

vibration
The rapid, back-and-forth movement of an object.

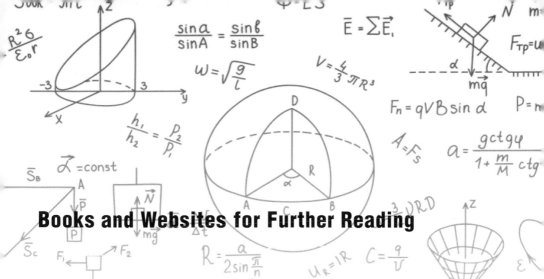

Books and Websites for Further Reading

Helget, N. *Full STEAM Baseball: Science, Technology, Engineering, Arts, and Mathematics of the Game.* Mankato, MN: Capstone Publishing, 2019.

Richards, Jon. *Baseball Superstars 2019: Top Players, Record Breakers, Facts & Stats.* New York: Sterling Publishing, 2019.

Savage, Jeff. *Baseball Super Stats.* Minneapolis, MN: Lerner Books, 2017.

Exploratorium Education:
https://www.exploratorium.edu/search/baseball

Smithsonian Institution Education:
https://www.si.edu/spotlight/baseball

Washington State University, *Ask Dr. Universe:*
https://askdruniverse.wsu.edu/?s=baseball

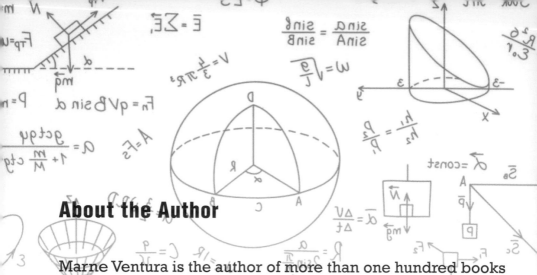

About the Author

Marne Ventura is the author of more than one hundred books for children. A former elementary school teacher, she holds a master's degree in reading and language development from the University of California. Marne's nonfiction titles cover a wide range of topics, including STEM, arts and crafts, food and cooking, biographies, health, and survival. Her fiction series, the *Worry Warriors*, tells the story of four brave kids who learn to conquer their fears. Marne and her husband live on the central coast of California.

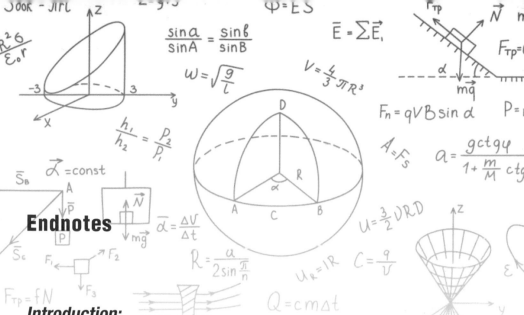

Endnotes

Introduction:

1. Gough, Christina. "World Series Average TV Viewership U.S. 2000-2019." *Statista*, 5 Nov. 2019, www.statista.com/statistics/235678/world-series-tv-viewership-in-the-united-states/.

Chapter 1:

1. "Newton's Laws of Motion." *NASA*, NASA, www.grc.nasa.gov/www/k-12/airplane/newton.html.
2. "Momentum." *Merriam-Webster*, Merriam-Webster, www.merriam-webster.com/dictionary/momentum.
3. "Putting Something On The Ball." *Science of Baseball*, www.exploratorium.edu/baseball/features/putting-something-on-the-ball.html.
4. Newman, Alan. "Alan Newman." *SciJourner*, 28 July 2016, www.scijourner.org/2012/03/02/the-science-behind-baseball/.
5. Westfall, Richard S. "Isaac Newton." *Encyclopædia Britannica*, Encyclopædia Britannica, Inc., 1 Jan. 2020, www.britannica.com/biography/Isaac-Newton.
6. "What Is a Curveball (CU)?: Glossary." *Major League Baseball*, m.mlb.com/glossary/pitch-types/curveball.

7.   Ward, Dennis. "Air Composition." *UCAR*, www.eo.ucar.edu/basics/wx_l_b_l.html.

8.   US Department of Commerce, et al. "Science Education - What Is Air Pressure?" *NDBC*, 8 Nov. 1996, www.ndbc.noaa.gov/educate/pressure.shtml.

9.   "Air Pressure and Humidity." *Infoplease*, Infoplease, www.infoplease.com/math-science/weather/air-pressure-and-humidity.

10. The Editors of Encyclopaedia Britannica. "Magnus Effect." *Encyclopædia Britannica*, Encyclopædia Britannica, Inc., 22 Dec. 2006, www.britannica.com/science/Magnus-effect.

11. "What Is a Curveball (CU)?: Glossary." *Major League Baseball*, m.mlb.com/glossary/pitch-types/curveball.

12. "Throw a Curve Ball." *The Free Dictionary*, Farlex, idioms.thefreedictionary.com/throw+a+curve+ball.

13. Chodosh, Sara. "The Physics behind a Baseball Bat's Sweet Spot." *Popular Science*, Popular Science, 30 Mar. 2018, www.popsci.com/baseball-bat-sweet-spot-science-physics/.

14. Coburn, Davin. "Baseball Physics: Anatomy of a Home Run." *Popular Mechanics*, Popular Mechanics, 14 Nov. 2017, www.popularmechanics.com/adventure/sports/a4569/4216783/.

15. "Yankees' Giancarlo Stanton Is the King of Exit Velocity." *SNY*, www.sny.tv/yankees/news/yankees-giancarlo-stanton-is-the-king-of-exit-velocity/284978664.

16. Landers, Chris. "Why Is It Called Tommy John Surgery." *MLB.com*, 27 Feb. 2019, www.mlb.com/cut4/why-is-it-called-tommy-john-surgery.

17. Sparks, Karen. "Frank Wilson Jobe." *Encyclopædia Britannica*, Encyclopædia Britannica, Inc., 12 July 2019, www.britannica.com/biography/Frank-Wilson-Jobe.

18. "Home Run Nutrition: What 5 Pro Baseball Players Eat." *Gazelle Nutrition Lab*, 28 Mar. 2019, gazellenutrition. com/what-5-pro-baseball-players-eat/.

19. "The D-Backs Will Be Serving a Churro Dog, Yes, a Churro Dog at Chase Field This Season." *MLB.com*, 4 Mar. 2015, www.mlb.com/cut4/the-d-backs-will-be-serving-a-churro-dog-yes-a-churro-dog-at-chase-field-this-season/c-111201476.

Chapter 2:

1. Jackson, Frank. "The Physics of Radar Guns." *The Hardball Times*, tht.fangraphs.com/the-physics-of-radar-guns/.

2. Woodford, Chris. "How Radar Works: Uses of Radar." *Explain That Stuff*, 14 Nov. 2019, www.explainthatstuff.com/radar.html.

3. See note 1 above.

4. Knoblor, Danny. "The Radar Gun Revolution." *Bleacher Report*, Bleacher Report, 3 Oct. 2017, bleacherreport. com/articles/2184581-the-radar-gun-revolution.

5. Sisney, Brock. "History of the Machine." *Morning Sun*, Morning Sun, 29 June 2013, www.morningsun.net/article/20130629/NEWS/306299967.

6. Eschenbach, Stephen, et al. "The First Pitching Machine." *The First Pitching Machine | Invention and Technology*, www.inventionandtech.com/content/first-pitching-machine-0?page=full.

7. See note 5 above.

8. "Master Pitching Machine." *D3 Technologies*, www. teamd3.com/about/success-stories/198/.

9. See note 6 above.

10. SportsLifer. "Ted Williams and the 10 Greatest Hitters That Ever Lived." *Bleacher Report*, Bleacher Report, 3

Oct. 2017, bleacherreport.com/articles/1240582-the-10-greatest-hitters-that-ever-lived.

11. "Tony Gwynn." *Tony Gwynn | Society for American Baseball Research*, sabr.org/bioproj/person/2236deb4.

12. Verducci, Tom. "The Technology Boom Is Fundamentally Altering Baseball." *Sports Illustrated*, 29 Mar. 2019, www.si.com/mlb/2019/03/29/technology-revolution-baseball-trackman-edgertronic-rapsodo.

13. "BASEBALL'S TECH REVOLUTION: Tribune Content Agency (August 14, 2019)." *Tribune Content Agency*, tribunecontentagency.com/article/baseballs-tech-revolution/.

14. Verducci, Tom. "The Technology Boom Is Fundamentally Altering Baseball." *Sports Illustrated*, 29 Mar. 2019, www.si.com/mlb/2019/03/29/technology-revolution-baseball-trackman-edgertronic-rapsodo.

15. The Editors of Encyclopaedia Britannica. "Tony Gwynn." *Encyclopædia Britannica*, Encyclopædia Britannica, Inc., 12 June 2019, www.britannica.com/biography/Tony-Gwynn.

16. "Virtual Reality." *Dictionary.com*, Dictionary.com, www.dictionary.com/browse/virtual-reality?s=t.

17. "The Use of Virtual and Augmented Reality in Baseball Training." *Baseball Reflections*, baseballreflections.com/2019/03/21/the-use-of-virtual-and-augmented-reality-in-baseball-training/.

18. Chew, Jonathan. "Why Major League Baseball Teams Are Turning to Virtual Reality." *Fortune*, Fortune, 29 Apr. 2016, fortune.com/2016/04/29/mlb-eon-sports-vr/.

19. "Jason Giambi for Project OPS by EON Sports VR." *EON Reality*, 31 Dec. 2019, eonreality.com/jason-giambi-for-project-ops-by-eon-sports-vr/.

20. "Virtual Reality Batting Practice Head-Set Is Dodgers' Real-Life Preparation Tool." *Los Angeles Times*, Los Angeles Times, 22 Aug. 2019, www.latimes.com/sports/dodgers/story/2019-08-22/virtual-reality-batting-goggles-headset-dodgers-baseball-chris-dan-odowd.

21. See note 18 above.

22. "Virtual Reality vs. Augmented Reality." *Augment News*, 6 Mar. 2017, www.augment.com/blog/virtual-reality-vs-augmented-reality/.

23. "MLB Takes AR to next Level for Fans at Ballpark." *MLB. com*, 10 Oct. 2017, www.mlb.com/news/mlb-to-use-augmented-reality-to-enhance-data-c258179374.

24. Jackson, Frank. "An Update On Wearable Baseball Technology." *The Hardball Times*, tht.fangraphs.com/an-update-on-wearable-technology/.

25. Goldman, Tom. "What's Up Those Baseball Sleeves? Lots Of Data, And Privacy Concerns." *NPR*, NPR, 30 Aug. 2017, www.npr.org/2017/08/30/547062884/whats-up-those-baseball-sleeves-lots-of-athletes-data-and-concerns-about-privacy.

26. Schelle, Charles. "Tommy John Surgery Prevention in a Sleeve at Motus Global's Bradenton Lab." *Bradenton*, Bradenton Herald, 13 July 2014, www.bradenton.com/news/business/technology/article34722444.html.

27. Berg, Ted, and Like. "How Wearable Technology Will Shape the Future of Hitting in MLB." *USA Today*, Gannett Satellite Information Network, 15 Jan. 2019, ftw.usatoday.com/2019/01/hitting-mlb-baseball-technology-driveline.

28. See note 24 above.

Chapter 3

1. Holtzman, Jerome, and Gilbert P. Laue. "History." *Encyclopædia Britannica*, Encyclopædia Britannica, Inc., 19 Nov. 2019, www.britannica.com/sports/baseball/History#ref782476.

2. "August 16, 1920: Ray Chapman Suffers Fatal Blow to His Skull on Pitch from Carl Mays." *August 16, 1920: Ray Chapman Suffers Fatal Blow to His Skull on Pitch from Carl Mays | Society for American Baseball Research*, sabr.org/gamesproj/game/august-16-1920-ray-chapman-suffers-fatal-blow-his-skull-pitch-carl-mays.

3. "B-R Bullpen." *BR Bullpen*, www.baseball-reference.com/bullpen/Batting_helmet.

4. Lukas, Paul. "The C-Flap Helmet Is Helping MLB Save Face." *ESPN*, ESPN Internet Ventures, 5 Apr. 2018, www.espn.com/mlb/story/_/id/23026863/the-mlb-c-flap-helmet-saving-faces-all-star.

5. "The Fascinating History Of The Baseball Glove." *ThePostGame.com*, 3 Feb. 9224, www.thepostgame.com/blog/throwback/201107/fascinating-history-baseball-glove.

6. Kutz, Steven. "Is This the Baseball Glove of the Future?" *MarketWatch*, 21 Oct. 2014, www.marketwatch.com/story/will-the-next-derek-jeter-wear-this-baseball-glove-2014-10-03.

7. See note 5 above.

8. "Baseball Glove Buyers Guide." *Baseball Gloves Buying Guide - Glove Webbing, Position, and Top Brands*, www.sportsunlimitedinc.com/baseball-glove-buyers-guide.html.

9. Stamp, Jimmy. "The Invention of the Baseball Mitt." *Smithsonian.com*, Smithsonian Institution, 16 July 2013, www.

smithsonianmag.com/arts-culture/the-invention-of-the-baseball-mitt-12799848/.

10. Soper, Taylor. "Diamond Kinetics Uses Embedded Sensors to Help MLB Batters and Pitchers Improve Performance." *GeekWire*, 7 Mar. 2018, www.geekwire.com/2018/diamond-kinetics-uses-embedded-sensors-help-mlb-batters-pitchers-improve-performance/.

11. Barrabi, Thomas. "The Baseball Bat Is Finally Getting A Tech Upgrade." *Fox Business*, Fox Business, 13 Apr. 2016, www.foxbusiness.com/features/the-baseball-bat-is-finally-getting-a-tech-upgrade.

12. Collins, Terry. "Swinging Away with the Smart Baseball Bat." *CNET*, CNET, 17 Apr. 2016, www.cnet.com/news/swinging-away-with-the-smart-baseball-bat/.

13. Axisa, Mike. "Mike Trout Is Using a 'Smart Bat' to Improve His Swing in Spring Training." *CBSSports.com*, 3 Mar. 2016, www.cbssports.com/mlb/news/mike-trout-is-using-a-smart-bat-to-improve-his-swing-in-spring-training/.

14. Darrow, Barb. "Amateur Baseball Players Get High-Tech Data To Sweeten Their Swings." *Fortune*, Fortune, 22 Mar. 2017, fortune.com/2017/03/22/sensors-improve-baseball-swings/.

15. Beck, Kellen. "This Smart Baseball Bat Can Teach a Complete Rookie How to Hit." *Mashable*, Mashable, 5 Apr. 2016, mashable.com/2016/04/05/zepp-smart-baseball-bat/.

16. No_Little_Plans. "We X-Rayed Some MLB Baseballs. Here's What We Found." *FiveThirtyEight*, FiveThirtyEight, 1 Mar. 2018, fivethirtyeight.com/features/juiced-baseballs/.

17. Peter, Josh. "These Scientists May Have Solved MLB's 'Juiced' Baseball Problem." *USA Today*, Gannett Satellite

Information Network, 5 Aug. 2019, www.usatoday.com/story/sports/mlb/2019/08/02/mlb-juiced-baseball-problem-home-run-rate/1869584001/.

18. Jackson, Nicholas. "The Complicated History of Baseball Stitching Machines." *The Atlantic*, Atlantic Media Company, 5 Jan. 2011, www.theatlantic.com/technology/archive/2010/10/the-complicated-history-of-baseball-stitching-machines/65274/.

19. Stamp, Jimmy. "A Brief History of the Baseball." *Smithsonian.com*, Smithsonian Institution, 28 June 2013, www.smithsonianmag.com/arts-culture/a-brief-history-of-the-baseball-3685086/.

20. "Properties of Baseball Bats." *Properties of Baseball Bats | Society for American Baseball Research*, sabr.org/research/properties-baseball-bats.

21. Stamp, Jimmy. "The Past and Future of the Baseball Bat." *Smithsonian.com*, Smithsonian Institution, 2 July 2013, www.smithsonianmag.com/arts-culture/the-past-and-future-of-the-baseball-bat-5618957/.

22. See note 20 above.

23. Kiger, Patrick J. "How Baseball Bats Work." *HowStuffWorks*, HowStuffWorks, 24 June 2013, entertainment.howstuffworks.com/baseball-bats1.htm.

24. "Baseball Bat." *How Products Are Made*, Encyclopedia.com, 23 Jan. 2020, www.encyclopedia.com/manufacturing/news-wires-white-papers-and-books/baseball-bat.

25. McDermott, Sean. "Take Me out on a Stretcher: The Dangers of Aluminum Bats in Baseball." *Bleacher Report*, Bleacher Report, 3 Oct. 2017, bleacherreport.com/articles/1149406-take-me-out-on-a-stretcher-the-case-against-aluminum-bats-in-baseball.

Chapter 4:

1. "Henry Chadwick." *Baseball Hall of Fame*, baseballhall. org/hall-of-famers/chadwick-henry.

2. "Henry Chadwick." *Henry Chadwick | Society for American Baseball Research*, sabr.org/bioproj/ person/436e570c.

3. "What Is a Batting Average (AVG)?: Glossary." *Major League Baseball*, m.mlb.com/glossary/standard-stats/ batting-average.

4. See note 1 above.

5. Augustyn, Adam. "Why Does 'K' Stand for a Strikeout in Baseball?" *Encyclopædia Britannica*, Encyclopædia Britannica, Inc., www.britannica.com/story/why-does-k-stand-for-a-strikeout-in-baseball.

6. "A Guide to Sabermetric Research." *A Guide to Sabermetric Research | Society for American Baseball Research*, sabr.org/sabermetrics/single-page.

7. Neyer, Rob. "Sabermetrics." *Encyclopædia Britannica*, Encyclopædia Britannica, Inc., 20 Aug. 2017, www.britannica.com/sports/sabermetrics.

8. "B-R Bullpen." *BR Bullpen*, www.baseball-reference. com/bullpen/Bill_James.

9. Fulton, Rob, and Rickey. "The Meaning of 'Moneyball'." *MinnPost*, 2 Feb. 2012, www.minnpost.com/ sports/2011/09/meaning-moneyball/.

10. See note 8 above.

11. "Bill James." *Bill James | Society for American Baseball Research*, sabr.org/about/bill-james.

12. Ebert, Roger. "Moneyball Movie Review & Film Summary (2011): Roger Ebert." *RogerEbert.com*, 21 Sept. 2011, www.rogerebert.com/reviews/moneyball-2011.

13. "Statistics." *Merriam-Webster*, Merriam-Webster, www. merriam-webster.com/dictionary/statistics.
14. "What Is a Batting Average (AVG)?: Glossary." *Major League Baseball*, m.mlb.com/glossary/standard-stats/ batting-average.
15. "What Is a Batting Average (AVG)?: Glossary." *Major League Baseball*, m.mlb.com/glossary/standard-stats/ batting-average.
16. Lokker, Brian. "History of MLB Batting Triple Crown Winners." *HowTheyPlay*, HowTheyPlay, 9 Jan. 2020, howtheyplay.com/team-sports/batting-triple-crown-winners-in-baseball.
17. "Every Triple Crown Winner in MLB History." *MLB.com*, 1 Nov. 2018, www.mlb.com/news/a-look-at-baseball-triple-crown-winners.
18. *ESPN*, ESPN Internet Ventures, www.espn.com/mlb/history/triplecrown.
19. "Standard Stats: Glossary." *Major League Baseball*, m.mlb.com/glossary/standard-stats.
20. "What Is a On-Base Percentage (OBP)?: Glossary." *Major League Baseball*, m.mlb.com/glossary/standard-stats/ on-base-percentage.
21. See note 19 above.
22. "What Is a Quality Start (QS)?: Glossary." *Major League Baseball*, m.mlb.com/glossary/standard-stats/quality-start.
23. "Official Rules." *Major League Baseball*, mlb.mlb.com/ mlb/fantasy/fb/info/official_rules.jsp.
24. Person. "A Look at the History of Baseball Cards." *Newsday*, Newsday, 22 Aug. 2010, www.newsday.com/sports/ baseball/a-brief-history-of-baseball-cards-1.2227938.
25. "B-R Bullpen." *BR Bullpen*, www.baseball-reference. com/bullpen/Baseball_card.

26. EBSCO Information Services, Inc. "The Application of Mathematics in Major League Baseball." *EBSCOpost Blog*, www.ebsco.com/blog/article/the-application-of-mathematics-in-major-league-baseball.

27. Sports, Written by Jobs In. "10 MLB Jobs for All Career Levels - Jobs In Sports." *JobsInSports Blog*, 31 Oct. 2018, www.jobsinsports.com/blog/2018/10/22/10-jobs-in-major-league-baseball-for-all-career-levels/.

28. Davoren, Julie. "Careers That Combine Math & Sports." *Chron.com*, 21 Nov. 2017, work.chron.com/careers-combine-math-sports-24662.html.

29. "New Class in Baseball Analytics Could Launch Careers." *Pomona Pitzer*, 24 Oct. 2019, www.sagehens.com/sports/bsb/2018-19/releases/2019102416pbfv.

30. See note 26 above.

31. "New Class in Baseball Analytics Could Launch Careers." *Pomona Pitzer*, 24 Oct. 2019, www.sagehens.com/sports/bsb/2018-19/releases/2019102416pbfv.

32. "MLB Sees Fan Growth across the Board in 2019." *MLB.com*, 30 Sept. 2019, www.mlb.com/news/mlb-increased-viewership-attendance-in-2019.

33. Thorne, Will. "TV Ratings: World Series Game 7 Draws 23 Million Viewers on Fox." *Variety*, 31 Oct. 2019, variety.com/2019/tv/news/tv-ratings-world-series-game-7-23-million-viewers-fox-1203389672/.